Your Days are Numbered in Calendar Math

by
Helene Silverman
Sheila Siderman

Cuisenaire Company of America, Inc.
10 Bank St., P.O. Box 5026
White Plains, N.Y. 10602-5026

Table of Contents

Introduction

Early math concepts and number computation are closely tied together in this book. Your Days are Numbered in Calendar Math is a sequential, eleven chapter book for teaching the calendar. It is also a helpful series of lessons for applying the concepts: before and after, direction, bar graphing, one and two-digit addition, multiples, and number patterns.

Each chapter of this book is a complete unit with its own teacher's notes and five different lessons. Intended to give children a lesson a week during the school year, the chapters increase in difficulty in terms of their calendar and mathematics content. One special feature throughout the book is the reference to calendars, past and future, to give interest and variation to each lesson. As part of each chapter, there are also instructions to children for creating their own current monthly calendar, and for using those calendars for the lessons.

You may begin this book at any time. Only the ''Just for Fun'' section of activities are oriented to specific months. Most of all, it is the intention of this book to give children a working understanding of the calendar, and to encourage the application of number skills.

Teacher Notes-Chapter One

Calendar Fill In

Objective: Relating the calendar to a numberline

Vocabulary: calendar, month, day, holidays, birthdays, first, last

Teaching the Lesson: Introduce the calendar as a broken numberline. Distribute a copy of the blank calendar in page 57 to each child. Assist the children as they write in the first and last day for the current month on their own calendars. Then help them to number the rest of their calendars. (Some children will need additional guide dates.) Assist the children to find the holidays, birthdays and special days for the month on their own calendars.

Number Find

Objective: Finding dates on the calendar

Vocabulary: date, today, today's date, week, day of the week

Teaching the Lesson: Use the calendars from the previous Calendar Fill In lesson to identify the month, day of the week and date. Call out numbers (dates) and ask the children to find them on their own calendars. (You may wish to use ordinal names.) Use a standard calendar to help the children to list the holidays and birthdays. Complete the activity on page 6.

Days of the Week

Objective: Learning the names of the days of the week

Vocabulary: days of the week, Sunday, Monday, Tuesday, Wednesday, Thursday, Friday, Saturday

Teaching the Lesson: Help the children to find the day and date on their own calendars. Identify the days of the week on the calendar. Demonstrate how to find the day for any given date. Guide the children to complete the activity on page 7.

Find My Neighbors

Objective: Learning to find the day before and the day after

Vocabulary: day before, day after

Teaching the Lesson: Use the July 1776 calendar on page 8. Help the children to locate ⏹10 on the calendar on page 8. Identify the day on which it occurs. Then, identify the day before and the day after. Assist the children as they complete the page. (While it is not necessary to use ordinal numbers, you may wish to use them in your oral work.)

Just for Fun Chapter Eleven contains a specific activity for each month. Find the activity for this month and duplicate copies for each child.

Number Find

Name_____

Use your personal calendar.

What month is it? _____

What day of the week is today? _____

What is today's date? _____

Look at your calendar.

Find 1. Draw ✓ in the box.

Find 2. Draw ✗ in the box.

Find 3. Draw ♡ in the box.

Find 4. Draw △ in the box.

Find 10. Draw ☺ in the box.

Find 13. Draw ⛵ in the box.

Find 17. Draw 🦋 in the box.

Find 19. Draw 🌳 in the box.

Find 20. Draw ◇ in the box.

Find 27. Draw ○ in the box.

Complete the charts.

BIRTHDAYS THIS MONTH	DATE

HOLIDAYS THIS MONTH	DATE

Days of the Week

Name_____

Today is_____

Use your calendar.
Name the 7 days of the week in order.
____Sunday____, _____, _____, _____,
_____, _____, _____.

Find all the Mondays on your calendar.
Write the dates. _____

Find all the Wednesdays.
Write the dates. _____

Find all the Saturdays.
Write the dates. _____

Circle the correct answer.
The first day of this month is
     Sunday Monday Tuesday Wednesday Thursday Friday Saturday.
The last day of this month is
     Sunday Monday Tuesday Wednesday Thursday Friday Saturday.

Write the day of the week for these dates this month.

    2 _____
    5 _____
    6 _____
    9 _____
  11 _____
  14 _____
  19 _____
  20 _____
  24 _____
  26 _____

| Sunday | Monday | Tuesday | Wednesday | Thursday | Friday | Saturday |

Find My Neighbors

Name_____

Today is_____

July 1776						
Sunday	Monday	Tuesday	Wednesday	Thursday	Friday	Saturday
	1	2	3	④	5	6
7	8	9	10	11	12	13
14	15	16	17	18	19	20
21	22	23	24	25	26	27
28	29	30	31			

July 4, 1776 — Signing of the Declaration of Independence.

Use the calendar above for this page.

Find ⑩ on the calendar. Circle it.
Draw boxes around ⑩ 's neighbors. Write the dates for:

     The day *after* ⑩ _____

     The day *before* ⑩ _____

Find ⑲ . Circle it.
Draw boxes around ⑲ 's neighbors. Write the dates for:

     The day *after* ⑲ _____

     The day *before* ⑲ _____

Write the dates for:

     The day before ③_____

     The day after ③_____

     The day before ㉙_____

     The day after ㉙_____

     The day before ㉛_____

     The day after ㉚_____

     The day before ㉗_____

     The day after ㉗_____

     The day before ⑭_____

     The day after ⑭_____

Your Days Are Numbered in Calendar Math   © 1980 Cuisenaire Company of America

Teacher Notes-Chapter Two

Calendar Fill In

Objective: Reviewing parts of the calendar

Vocabulary: month, days, weeks, date, Sunday, Monday, Tuesday, Wednesday, Thursday, Friday, Saturday, first, last

Teaching the Lesson: Introduce the calendar for this month using a standard calendar. Help the children to find the name of the month, the first and last day, birthdays, holidays, and special class days. Distribute copies of the blank calendar on page 58. Help the children to fill in the calendar for this month and to identify the day and date. Provide oral practice reading dates on the calendar. (While it is not necessary to use ordinal numbers, you may wish to use them in oral work.)

Finding Dates

Objective: Identifying dates on a calendar

Vocabulary: today, tomorrow, yesterday, days from today, one week, next week, coming week, days between

Teaching the Lesson: Use a standard calendar and the calendar from the previous Calendar Fill In lesson. Help the children to identify the month, day of the week, and date. Identify the day before as yesterday and the day after as tomorrow. Select pairs of dates and help the children to find the dates between. Read the dates for each week of this month with the children. Assist the children as they complete the activity on page 10.

The Dating Game

Objective: Reviewing position on a calendar; practicing one and two digit addition

Vocabulary: lines, score

Teaching the Lesson: Use the calendar from the previous Calendar Fill In lesson. Demonstrate the game on page 11. Assist the children as they play independently. Pair the children so that they can play rounds.

Arrow Addition

Objective: Practicing one and two digit addition on the calendar

Vocabulary: arrows, boxes, sum

Teaching the Lesson: Use the calendar on page 12. Help the children to locate the four dates on the calendar forming the box:

21	22
28	29

Assist them to complete the activity on page 12. (You may wish to extend the activity by asking the children to make addition boxes on their own calendars.) Help the children to notice that the arrow sums are equal.

Just for Fun

Chapter Eleven contains a specific activity for every month. Find this month's activity and duplicate copies for the class.

Finding Dates

Name _____

Today is _____

Tomorrow will be _____

Yesterday was _____

Use your calendar. Circle today's date.
Find the dates for this week starting with today.
Write them in order.

_____

Find the dates for next week. Begin with Sunday.
Write them in order.

_____

What dates come between ⑦ and ⑩ ? _____ _____
What dates come between ⑮ and ⑱ ? _____ _____
What dates come between ㉒ and ㉕ ? _____ _____

Write these dates:

2 days from today _____

5 days from today _____

7 days from today _____

14 days from today _____

1 day ago _____

2 days ago _____

7 days ago _____

Write the day of the week for these dates this month.

1 _____

3 _____

8 _____

19 _____

28 _____

11 _____

22 _____

4 _____

18 _____

10 _____

The Dating Game

Name _____

Today is _____

Tomorrow will be _____

Yesterday was _____

Play the Dating Game on your calendar.
You will need a counter or button.

First play by yourself.
Toss the counter onto the calendar.
Write the dates in the boxes the counter touches.
Find their sum.
Toss and add two more times.
Find the sum of the three tosses.

MAY 1991						
Sunday	Monday	Tuesday	Wednesday	Thursday	Friday	Saturday
				1	2	3	4
5	6	7	8	9	10	11
12	13	14	15	16	17	18
19	20	21	22	23	24	25
26	27	28	29	30	31	

Toss	Dates Touched	Sum
1		
2		
3		
		Sum of 3 Tosses

Now play with a friend.
Take turns. Keep score.
Keep adding your score after each turn.
The first person to reach 100 wins.

Toss	Dates Touched	Sum	Running Total

Arrow Addition

Name _____

Today is _____

Tomorrow will be _____

Yesterday was _____

January 2000						
Sunday	Monday	Tuesday	Wednesday	Thursday	Friday	Saturday
						1
2	3	4	5	6	7	8
9	10	11	12	13	14	15
16	17	18	19	20	21	22
23	24	25	26	27	28	29
30	31					

Pick 4 touching boxes on a calendar.
Follow the arrows and add.

$$\begin{array}{r} 22 \\ +28 \end{array} \qquad \begin{array}{r} 21 \\ +29 \end{array}$$

Try some more. Follow the arrows and add.
Look at the sums.
What do you notice?

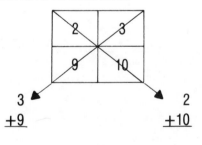

$$\begin{array}{r} 3 \\ +9 \end{array} \qquad \begin{array}{r} 2 \\ +10 \end{array}$$

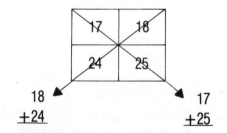

$$\begin{array}{r} 18 \\ +24 \end{array} \qquad \begin{array}{r} 17 \\ +25 \end{array}$$

Use the calendar above to complete these boxes.
Then draw the arrows and add.
Look at the sums each time.
What do you notice?

18	19

	24
	31

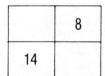

Your Days Are Numbered in Calendar Math © 1980 Cuisenaire Company of America

Teacher Notes-Chapter Three

Calendar Fill In

Objective: Naming the days of the week

Vocabulary: today, tomorrow, yesterday, Sunday, Monday, Tuesday, Wednesday, Thursday, Friday, Saturday

Teaching the Lesson: Introduce the calendar for this month using a standard calendar. Help the children to identify the name of the month, the first and last days, holidays, birthdays, and special class days. Guide the children as they fill in the calendar for the month using the blank calendar on page 59. A "Line-Up of Days" is provided to help the children to write the names of the days of the week. Provide oral practice finding the days on which different dates fall.

Finding Days

Objective: Finding days of the week

Vocabulary: days between, days from today, days of the week, weeks ago, days ago

Teaching the Lesson: Review the names of the days of the week with the children. Begin with any day and repeat the sequence in order. Guide the children to complete the activity on page 14.

Shortcuts

Objective: Writing abbreviations for the names of the days of the week

Vocabulary: Sun., Mon., Tues., Wed., Thurs., Fri., Sat.

Teaching the Lesson: Discuss the practice of using short forms for the names of the days of the week since they are so long and all have the same ending. Point out that most of the abbreviations just leave out "day" from the spelling. (The exceptions are Wednesday and Saturday.) Guide the children to complete the activity on page 15.

Points for Shortcuts

Objective: Practicing spelling of the abbreviations for the days of the week; practicing one and two digit addition

Vocabulary: list, points, most, least

Teaching the Lesson: Review the names of the days of the week and the spelling of the abbreviations. Introduce the point list on page 16. Guide the children to complete the activity. (You may wish to extend the activity to the spelling of the full names of the days of the week.)

Just for Fun

Chapter Eleven contains a specific activity for every month. Find the activity for this month and duplicate copies for each child.

Finding Days

Name _____

Today is _____

Tomorrow will be _____

Yesterday was _____

There are 7 days in a week.
Their names are:

| Sunday | Monday | Tuesday | Wednesday | Thursday | Friday | Saturday |

Write the day of the week:

Today _____

Tomorrow _____

Yesterday _____

2 days from today _____

2 days ago _____

A week from today _____

A week ago _____

What days come between? Complete:

Monday, _____, Wednesday

Friday, _____, Sunday

Saturday, _____, Monday

Thursday, _____, _____, Sunday

Tuesday, _____, _____,

_____, _____,

_____, _____, Tuesday

Wednesday, _____, _____,

_____, _____, Monday

Shortcuts

Name_____

Today is _____

Tomorrow will be _____

Yesterday was _____

You can use shortcuts to write the days of the week.

DAY	SHORTCUT
Sunday	Sun.
Monday	Mon.
Tuesday	Tues.
Wednesday	Wed.
Thursday	Thurs.
Friday	Fri.
Saturday	Sat.

Match the day to its shortcut.

Tuesday	Sun.
Saturday	Sat.
Wednesday	Tues.
Friday	Mon.
Sunday	Fri.
Thursday	Wed.
Monday	Thurs.

Complete the chart.

SHORTCUT	DAY
Wed.	
Sat.	
Mon.	
Thurs.	
Tues.	
Fri.	
Sun.	

Sun.	Mon.	Tues.	Wed.	Thurs.	Fri.	Sat.

Points for Shortcuts

Name _____

Today is _____

Tomorrow will be _____

Yesterday was _____

Look at the letter list.
Each letter gets points.
How many points for each shortcut?

SHORTCUT	POINTS
Sun.	19 + 21 + 14 = 54
Mon.	
Tues.	
Wed.	
Thurs.	
Fri.	
Sat.	

Which shortcut has the most points? _____

Which shortcut has the fewest points? _____

Find the difference. _____

⭐ SUPER How many points for this month's name?

_____

How many points for your name?

_____

LETTER LIST	
A =	1
B =	2
C =	3
D =	4
E =	5
F =	6
G =	7
H =	8
I =	9
J =	10
K =	11
L =	12
M =	13
N =	14
O =	15
P =	16
Q =	17
R =	18
S =	19
T =	20
U =	21
V =	22
W =	23
X =	24
Y =	25
Z =	26

Teacher Notes-Chapter Four

Calendar Fill In

Objective: Using the abbreviated form for the days of the week

Vocabulary: Sun., Mon., Tues., Wed., Thurs., Fri., Sat.

Teaching the Lesson: Introduce the calendar for this month using a standard calendar. Help the children to identify the name of the month, the first and last days, birthdays, holidays, and special class days. Guide the children as they fill in the calendar for the month on the blank calendar on page 60. Emphasize writing the short forms for the names of the days of the week. Provide oral practice finding the names of the days and dates of birthdays, holidays, and special class days.

Odd and Even

Objective: Finding odd and even numbers

Vocabulary: odd, even, twos

Teaching the Lesson: Help the children to find $\boxed{2}$ on the calendar developed in the previous Calendar Fill In lesson. Guide them as they count by twos, circling every number they say. Assist them as they complete the activity on page 18. (While it is not necessary to use ordinal numbers, you may wish to use them.)

Threes

Objective: Counting by threes on the calendar

Vocabulary: threes

Teaching the Lesson: Help the children to find $\boxed{3}$ on the calendar for the month. Teach the children to count by threes, listing each number they say. Help the children to count by threes on the calendar on page 19 and to circle each number they say. Guide the children to complete the activity on page 19. (While it is not necessary to use ordinal numbers, you may wish to use them.)

Fours

Objective: Counting by fours on the calendar

Vocabulary: fours

Teaching the Lesson: Assist the children to count by fours, making a list of the numbers they say. Help them to count by fours on the calendar on page 20 and to circle the numbers they say. Guide them to complete the activity on page 20. (While it is not necessary to use ordinal numbers, you may wish to use them.)

Just for Fun Chapter Eleven contains a specific activity for every month. Find the activity for this month and duplicate copies for each child.

Odd and Even

Name _____

Today is _____

Tomorrow will be _____

Yesterday was _____

Find ② on your calendar. Color it red.
Count by 2's.
Circle each number you say.
Color in all the boxes with circles.
Write all the numbers you colored.

2, 4, _____

_____

These numbers are *even* numbers.

Write all the numbers you didn't color.

_____

_____

These numbers are *odd* numbers.

Find each date on your calendar.

Complete the chart.

SEARCH AND FIND	DATE	ODD OR EVEN?
Today		
Tomorrow		
Yesterday		
Next Friday		
Last Tuesday		
Special Days		

EVEN | 2 | 4 | 6 | 8 | 10 | 12 | 14 | 16 | 18 | 20 | 22 | 24 | 26 | 28 | 30 |

ODD | 1 | 3 | 5 | 7 | 9 | 11 | 13 | 15 | 17 | 19 | 21 | 23 | 25 | 27 | 29 | 31 |

Threes

Name_____

Today is _____

Tomorrow will be _____

Yesterday was _____

AUGUST 1999						
Sunday	Monday	Tuesday	Wednesday	Thursday	Friday	Saturday
1	2	3	4	5	6	7
8	9	10	11	12	13	14
15	16	17	18	19	20	21
22	23	24	25	26	27	28
29	30	31				

Use the calendar above. Find ③. Circle it.
Count by 3's. Circle each number you count.
Write the numbers you circled.

3, 6, _____

Use the calendar above to solve the riddles.

I am a Wednesday.
You circled me.
Who am I? _____

I am a Friday.
Your circled me in the
 first week.
Who am I? _____

I am a Monday.
I am less than 10.
You did not circle me.
Who am I? _____

I am a Tuesday.
I am more than 10.
You circled me.
Who am I?

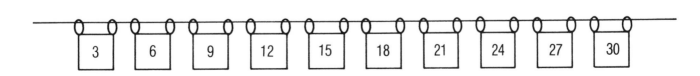

Fours

Name _____

Today is _____

Tomorrow will be _____

Yesterday was _____

NOVEMBER 1985						
Sunday	Monday	Tuesday	Wednesday	Thursday	Friday	Saturday
					1	2
3	4	5	6	7	8	9
10	11	12	13	14	15	16
17	18	19	20	21	22	23
24	25	26	27	28	29	30

Use the calendar above.

Find ④. Circle it.

Count by 4's. Circle each number you count.

Write the numbers you circled.

 _____

Use the calendar above to solve the riddles.

I am a Saturday.
You circled me.
Who am I? _____

I am a Tuesday.
You did not circle me.
The sum of my digits is 10.
Who am I? _____

I am a Wednesday.
You did not circle me.
I am less than 10.
Who am I? _____

You circled me.
I am more than 10.
The sum of my digits is 2.
Who am I? _____

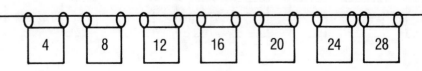

Your Days Are Numbered in Calendar Math

Teacher Notes-Chapter Five

Calendar Fill In

Objective: Finding sequences on the calendar

Vocabulary: odd, even, twos, threes, fours

Teaching the Lesson: Introduce the calendar for this month using a standard calendar. Help the children to identify the name of the month, the first and last days, the holidays, birthdays, and special class days. Help the children to fill in the calendar on page 61. Review sequences on the calendar, counting by twos, threes, and fours.

Let's Take a Walk

Objective: Using positional language on a calendar

Vocabulary: up, down, right, left

Teaching the Lesson: Help the children to find $\boxed{10}$ on the calendar on page 22. Teach the children to move one right from the $\boxed{10}$. Then move one left, one right, one up, and one down. Assist the children with the activity on page 22.

Find the Day

Objective: Reviewing names of the days of the week

Vocabulary: Sunday, Monday, Tuesday, Wednesday, Thursday, Friday, Saturday

Teaching the Lesson: Review the names of the days of the week. Help the children to unscramble the sequences on page 23 using the Line-Up of Days as a reference.

Digit Doings

Objective: Reading digits; practicing with single digit addition

Vocabulary: digits, add, sum

Teaching the Lesson: Help the children to find dates on the calendars developed in the previous Calendar Fill In lesson. Talk about the digits in a date. For example, the digits in $\boxed{21}$ are $\boxed{2}$ and $\boxed{1}$. The sum of the digits is $2+1$ or 3. Help the students orally locate some dates and find the digits and their sum. Guide them as they complete the activity on page 24.

Just for Fun

Chapter Eleven contains a specific activity for each month. Find the activity for this month and duplicate copies for each child.

Let's Take a Walk

Name _____

Today is _____

Tomorrow will be _____

Yesterday was _____

MARCH 2020						
Sunday	Monday	Tuesday	Wednesday	Thursday	Friday	Saturday
1	2	3	4	5	6	7
8	9	10	11	12	13	14
15	16	17	18	19	20	21
22	23	24	25	26	27	28
29	30	31				

Use the calendar above.

Let's take a walk on the calendar.

These are the ways you can walk.

Up ↑   Down ↓   Right →   Left ←

BEGIN AT	GO	END
16	Right 1	17
26	Up 2	
7	Left 3	
12	Down 2	

Now try these.

Begin at 22. Go right 2. Where are you? _____

    Then go up 3. Where are you? _____

    Then go left 2. Where are you? _____

    Then go down 1. Where are you? _____

Begin at 7. Go down 1. _____

    Then go down 2. _____

Begin at 28. Then go left 3. _____

    Then go right 1. _____

    Then go up 2. _____

SUPER ★ S

Your Days Are Numbered in Calendar Math   © 1980 Cuisenaire Company of America

Find the Day

Name_____

Today is _____

Tomorrow will be _____

Yesterday was _____

The names of the 7 days are hidden below.
The shortcuts are also hidden.
Circle the days and their shortcuts.

M	O	N	D	A	Y	F	A	S	N
O	S	T	S	M	A	R	C	U	W
F	A	H	M	O	N	I	H	N	E
I	T	H	U	R	S	D	A	Y	D
S	U	N	D	A	Y	A	H	L	N
F	R	I	S	N	O	Y	E	I	E
E	D	O	R	M	T	H	U	R	S
T	A	S	D	V	U	W	A	O	D
Y	Y	A	N	T	E	E	R	S	A
W	T	T	U	E	S	D	A	Y	Y

Use this checklist to help you find the days and shortcuts.

Monday	Mon.
Tuesday	Tues.
Wednesday	Wed.
Thursday	Thurs.
Friday	Fri.
Saturday	Sat.
Sunday	Sun.

Digit Doings

Name _____

Today is _____

Tomorrow will be _____

Yesterday was _____

Use your calendar to complete the chart.

FIND THE DATES.	SMALLEST NUMBER?	LARGEST NUMBER?
All dates with 3: 3 13	3	
All dates with 4:		
All dates with 5:		

Look at 14. The digits are 1 and 4.
The digit sum is 1 + 4, or 5.
Find other digit sums.
Complete the chart.

DATE	DIGITS	DIGIT SUM
23	2 + 3	5
18		
26		
11		
10		

SUPER ★ S

Write today's date. _____

What is the digit sum? _____

List 5 dates with larger digit sums. _____

Which date has the largest digit sum? _____ What is the sum? _____

List 5 dates with smaller digit sums. _____

Which date has the smallest digit sum? _____ What is the sum? _____

Teacher Notes-Chapter Six

Calendar Fill In

Objective: Reviewing calendar language

Vocabulary: day, date, day before, day after, event

Teaching the Lesson: Introduce the calendar for this month using a standard calendar. Ask the children to identify the name of the month, the first and last days, holidays, birthdays, and special class days. Distribute a copy of the blank calendar on page 62 to each child. Help the children to make a list of holidays, birthdays, and special class days. Guide them as they fill in the calendar for the month. (It may be helpful at this time to find the date, day, day before and day after each day on the list.)

Months of the Year

Objective: Naming the months

Vocabulary: January, February, March, April, May, June, July, August, September, October, November, December, next month, last month

Teaching the Lesson: Introduce the names of the months of the year in order. Then begin with any month and name the next twelve months in order. Ask the children to begin with the current month and name the months in order. Guide the children to complete the activity on page 26.

Monthly Shortcuts

Objective: Writing abbreviations for the months

Vocabulary: abbreviation, Jan., Feb., Mar., Apr., May, June, July, Aug., Sept., Oct., Nov., Dec.

Teaching the Lesson: Review the Monthly Line-Up. Introduce the abbreviated forms for the names of the months. Complete the activity on page 27.

Hidden Months

Objective: Recognizing the names of the months

Vocabulary: January, February, March, April, May, June, July, August, September, October, November, December

Teaching the Lesson: Review the names of the months. Guide the children to complete the activity on page 28.

Just for Fun

Chapter Eleven contains a specific activity for every month. Find the activity for this month and duplicate copies for each child.

Months of the Year

Name _____

Today is _____

Tomorrow will be _____

Yesterday was _____

There are 12 months in a year.
Their names are:

January   February   March
April   May   June   July
August   September   October   November   December

Write the name of the month:

Today _____

Yesterday _____

A week ago _____

Last month _____

Next month _____

Two months from today _____

Six months from today _____

A year ago _____

A year from today _____

My birthday _____

What month is this? _____

Write the next 12 months in order.

_____

_____

_____

Monthly Shortcuts

Name _____

Today is _____

Tomorrow will be _____

Yesterday was _____

You can use the shortcuts to write the months.

MONTH	SHORTCUT	MONTH	SHORTCUT
January	Jan.	**July**	July
February	Feb.	**Aug**ust	Aug.
March	Mar.	**Sept**ember	Sept.
April	Apr.	**Oct**ober	Oct.
May	May	**Nov**ember	Nov.
June	June	**Dec**ember	Dec.

Which months do not have abbreviations?

_____ _____ _____ _____

Complete the names of the months.

Ja____uar____   ____ ____ly

Fe____ ____ ____ary   Au____ ____s____

M____ ____ ____h   S____ ____tem____er

A____ ____i____   Oct____ ____ ____ ____

____ay   ____ ____ ____ember

____ ____ne   ____ ____ ____ ____m____ ____

Write the months in order using abbreviations:

_____ _____ _____ _____ _____ _____

_____ _____ _____ _____ _____ _____

Complete the abbreviations for the months below:

Your Days Are Numbered in Calendar Math © 1980 Cuisenaire Company of America

27

Hidden Months

Name _____

Today is _____

Tomorrow will be _____

Yesterday was _____

The names of the 12 months are hidden below.
Find them.

F	E	B	R	U	A	R	Y	A	E
N	O	V	E	M	B	E	R	P	D
O	C	T	O	B	E	R	J	R	E
V	A	H	T	V	L	W	M	I	C
A	I	R	A	G	E	F	A	L	E
P	S	J	U	L	Y	M	R	P	M
R	M	U	G	D	S	A	C	C	B
J	A	N	U	A	R	Y	H	P	E
A	R	E	S	J	K	O	B	N	R
S	E	P	T	E	M	B	E	R	B

Use this checklist to help you find the months.

January	July
February	August
March	September
April	October
May	November
June	December

   Your Days Are Numbered in Calendar Math   © 1980 Cuisenaire Company of America

Teacher Notes-Chapter Seven

Calendar Fill In

Objective: Figuring days until an event

Vocabulary: days until

Teaching the Lesson: Introduce the calendar for this month using a standard calendar. Help the children to identify the name of the month, the first and last days, holidays, birthdays, and special class days. Duplicate and distribute copies of the blank calendar on page 58. Guide the children as they fill in the blank calendar. Make a list of special days. Help the children to find the number of days until each special day.

Fives and Sixes

Objective: Counting by fives and sixes on the calendar

Vocabulary: fives, sixes and multiples

Teaching the Lesson: Help the children to find $\boxed{5}$ on the calendar on page 30. Orally count by fives as a group. Write the multiples of 5, i.e. 5, 10, 15, 20, 25, 30. Help the children to find the multiples for 6, i.e. 6, 12, 18, 24, 30. Help the children to complete the activity on page 30 using the calendar on the page.

Days in a Month

Objective: Learning the number of days in a month

Vocabulary: Leap year, all the rest, except

Teaching the Lesson: Review the names of the months with the children. Set up a Monthly Line Up. Teach the poem, "Thirty Days Has September". Help the children to complete the activity on page 31.

Birthdays

Objective: Collecting and presenting information

Vocabulary: chart, bar graph, most, least

Teaching the Lesson: Review the Monthly Line Up with the class. Ask each child to write down the name of the month in which his birthday occurs. Prepare a large chart like the one on page 32. Collect the birthday information for members of the class. Help the children to complete the activity on page 32.

Just for Fun

Chapter Eleven contains a specific activity for every month. Find the activity for this month and duplicate copies for each child.

Fives and Sixes

Name_____

Today is _____

Tomorrow will be _____

Yesterday was _____

MAY 1991						
Sunday	Monday	Tuesday	Wednesday	Thursday	Friday	Saturday
			1	2	3	4
5	6	7	8	9	10	11
12	13	14	15	16	17	18
19	20	21	22	23	24	25
26	27	28	29	30	31	

Use the calendar above.
Find 5. Circle it in pen.
Count by 5's. Circle each number you count.
Which dates are multiples of 5? _____

Find 6. Circle it in pencil.
Count by 6's. Circle each number you count.
Which dates are multiples of 6?_____

Complete the line-ups.

| 5 | 10 | | | | |

| 6 | 12 | | | |

Which date is a multiple of 5 and 6? _____

Use the calendar to solve the riddles.

I am a multiple of 5.
I fall on a Friday.
Who am I? _____

I am a multiple of 6.
I fall on a Saturday.
Who am I? _____

I am a multiple of 5
    and a multiple of 6.
Who am I? _____
    What day do I fall on?_____

I am a Friday.
I am not a multiple of 5.
I am not a multiple of 6.
I am more than 18.
Who am I? _____

  Your Days Are Numbered in Calendar Math   © 1980 Cuisenaire Company of America

Days in a Month

Name _____

Today is _____

Tomorrow will be _____

Yesterday was _____

Thirty days has September,
April, June, and November.
All the rest have 31,
Except February which has 28 alone
Until leap year gives it 29.

Which months have 30 days? _____

_____

Which months have 31 days? _____

_____

Which month has 28 days except when leap year gives it 29? _____

How many days for each month?

January _____	February _____	March _____
April _____	May _____	June _____
July _____	August _____	September _____
October _____	November _____	December _____

Complete:

Today is June 30. Tomorrow is _____ .

Today is May 30. Tomorrow is _____ .

Today is February 29. Tomorrow is _____ .

Today is January 30. Tomorrow is _____ .

Today is December 31. Tomorrow is _____ .

Today is June 1. Yesterday was _____ .

Today is November 1. Yesterday was _____ .

Today is February 1. Yesterday was _____ .

Today is April 1. Yesterday was _____ .

Today is January 1. Yesterday was _____ .

Birthdays

Name _____

Today is _____

Tomorrow will be _____

Yesterday was _____

Find out the birthdays of all the children in your class.
Write each name next to the month.
How many birthdays in each month?
Make a bar graph.

MONTH	STUDENTS	HOW MANY?
January		
February		
March		
April		
May		
June		
July		
August		
September		
October		
November		
December		

CLASS BIRTHDAYS

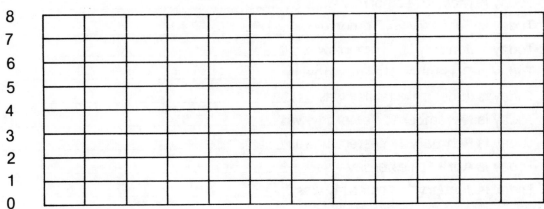

Your Days Are Numbered in Calendar Math © 1980 Cuisenaire Company of America

Teacher Notes-Chapter Eight

Calendar Fill In

Objective: Reviewing counting patterns

Vocabulary: multiples, odds, evens, twos, threes, fours, fives, sixes

Teaching the Lesson: Introduce the calendar for the month using a standard calendar. Help the children to identify the name of the month, the first and last days, the holidays, birthdays, and special class days. Duplicate and distribute copies of the blank calendar on page 63 to each child. Help the children to complete the Event Classification, i.e. whether the day is odd or even, a multiple of two, three, four, five, or six.

Weekdays and Weekends

Objective: Identifying weekdays and weekends

Vocabulary: weekends, weekdays

Teaching the Lesson: Use the monthly calendar developed in the previous Calendar Fill In lesson. Help the children to find the Saturdays and Sundays on their calendars. Teach the word "weekend". Help the children to find the weekends for this month. Help the children to identify the remaining days as "schooldays" or "weekdays". Guide them to find the weekdays for the current month. Help them to identify whether the events for this month occur on weekends or weekdays. Guide them to complete the activity on page 34.

Sevens

Objective: Finding multiples of seven

Vocabulary: sevens, multiple, less than, more than

Teaching the Lesson: Help the children to find 7 on their monthly calendars. Guide them to count orally by sevens, i.e. 7, 14, 21, 28. Write the multiples of seven on the blackboard. Guide the children to complete the activity on page 35.

Week Before and Week After

Objective: Recognizing patterns

Vocabulary: week before, week after, pattern

Teaching the Lesson: Review the number of days in a week. Help the children to identify the weeks on their calendars developed in the previous Calendar Fill In lesson. Guide them to orally name the days in each week. Help them to complete the activity on page 36.

Just for Fun

Chapter Eleven contains a specific activity for every month. Find the activity for this month and duplicate copies for each child.

Weekdays and Weekends

Name _____

Today is _____

Tomorrow will be _____

Yesterday was _____

Use your calendar.
Find all the Saturdays.
Circle the dates.
Find all the Sundays.
Draw boxes around the dates.

Which dates are Saturdays? _____

Which dates are Sundays? _____

Saturday and Sunday together are called a weekend.
Write pairs of dates for the weekends this month.

_____ & _____, _____ & _____, _____ & _____, _____ & _____

The days left are called weekdays.
Monday, Tuesday, Wednesday, Thursday, and Friday are weekdays.

Write the dates of the weekdays this month.

_____ _____ _____ _____ _____

_____ _____ _____ _____ _____

_____ _____ _____ _____ _____

_____ _____ _____ _____ _____

_____ _____ _____ _____ _____

_____ _____ _____ _____ _____

Your Days Are Numbered in Calendar Math   © 1980 Cuisenaire Company of America

Sevens

Today is _____

Tomorrow will be _____

Yesterday was _____

JANUARY 1996						
Sunday	Monday	Tuesday	Wednesday	Thursday	Friday	Saturday
	1	2	3	4	5	6
7	8	9	10	11	12	13
14	15	16	17	18	19	20
21	22	23	24	25	26	27
28	29	30	31			

Use the calendar above.

Find 7. Circle it.

Count by 7's.

Circle each number you count.

Which dates are multiples of 7?_____

Do they all fall on the same day? _____ Which day? _____

Complete the line-up for 7.

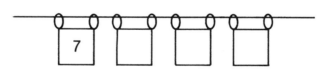

Use the calendar above to solve the riddles.

I am a multiple of 7.
I am less than 10.
Who am I? _____

I am a multiple of 7.
I am more than 12.
I am less than 15.
Who am I? _____

I am a multiple of 7.
The sum of my digits is 10.
Who am I? _____

I am more than 25.
I fall on a weekend.
I am not a multiple of 7.
Who am I? _____

Week Before and Week After

Name_____

Today is _____

Tomorrow will be _____

Yesterday was _____

MARCH 2002						
Sunday	Monday	Tuesday	Wednesday	Thursday	Friday	Saturday
					1	2
③	4	5	6	7	8	9
☐10	11	12	13	14	15	16
⑰	18	19	20	21	22	23
24	25	26	27	28	29	30
31						

Use the calendar above.

Look at Sunday, March ☐10 .

One week after is Sunday, and the date is ⑰ .

One week before is Sunday, and the date is ③ .

Complete the chart for March, 2002.

DATE	DAY	ONE WEEK AFTER	ONE WEEK BEFORE
12	Tuesday	19	5
15			
22			
23			
19			
3			
29			

A special pattern:

Look at the Thursdays.

What are the dates?_____

What are these numbers? _____

     Your Days Are Numbered in Calendar Math   © 1980 Cuisenaire Company of America

Teacher Notes-Chapter Nine

Calendar Fill In

Objective: Using symbols

Vocabulary: weather, rainy, sunny, cloudy

Teaching the Lesson: Introduce the calendar for this month using a standard calendar. Help the children to identify the name of the month, the first and last days, holidays, birthdays, and special class days. Duplicate the blank calendar on page 64 and distribute copies to each child. Introduce the symbols for rainy, sunny and cloudy. Help the children to fill in the calendar for the month and to begin to record the daily weather. (You may wish to use additional symbols.)

Eights and Nines

Objective: Finding multiples of eight and nine

Vocabulary: multiples, eights, nines

Teaching the Lesson: Help the children to count by eight on the calendar developed during the previous Calendar Fill In lesson. Make a Line-Up for the multiples of eight, i.e. 8, 16, 24. Help the children to find the multiples of nine and make a Line-Up for multiples of nine. Guide the children to complete the activity on page 38.

Weather Report

Objective: Recording and interpreting information

Vocabulary: rain, weather, weekend, weekday

Teaching the Lesson: Review the weather symbol for rain. Help the children to record the weather for the day using the calendar developed from the previous Calendar Fill In lesson. Guide the children to complete the activity on page 39.

Seasons

Objective: Identifying the seasons

Vocabulary: seasons, summer, fall, winter, spring

Teaching the Lesson: Duplicate a list of the names of the members of the class or make a chart. Write the birthday for each child. Help the children to use the list or chart to complete the activity on page 40. (You may wish to also make a graph of the number of birthdays for each season.)

Just for Fun

Chapter Eleven contains a specific activity for every month. Find the activity for this month and duplicate copies for each child.

Eights and Nines

Name _____

Today is _____

Tomorrow will be _____

Yesterday was _____

Use your own calendar.
Find ⑧ Circle it.
Count by 8. Circle each number you count.
Which dates are multiples of 8? _____
Complete the line-up for 8.

Find 9. Draw a box around it.
Count by 9. Draw a box around each number you count.
Which dates are multiples of 9? _____
Complete the line-up for 9.

List the multiples of 8 and 9 on the calendar.
Find the sums.

Multiples of 8	Multiples of 9
_____	_____
_____	_____
+ _____	+ _____
sum	sum

Use your calendar to solve these riddles.

I am a multiple of 8.
I am less than 10.
Who am I? _____

I am a multiple of 9.
I am greater than 20.
Who am I? _____

I am a multiple of 8.
The sum of my digits is 7.
Who am I? _____

I am a multiple of 9.
I am greater than 10
    and less than 20.
Who am I? _____

   Your Days Are Numbered in Calendar Math   © 1980 Cuisenaire Company of America

Weather Report

Name _____

Today is _____

Tomorrow will be _____

Yesterday was _____

APRIL 1990

Sunday	Monday	Tuesday	Wednesday	Thursday	Friday	Saturday
1 Sunny	2 Sunny	3 Partly Cloudy	4 Partly Cloudy	5 Rainy	6 Rainy	7 Sunny
8 Partly Cloudy	9 Partly Cloudy	10 Partly Cloudy	11 Sunny	12 Sunny	13 Sunny	14 Sunny
15 Rainy	16 Rainy	17 Rainy	18 Rainy	19 Partly Cloudy	20 Partly Cloudy	21 Partly Cloudy
22 Sunny	23 Sunny	24 Sunny	25 Rainy	26 Sunny	27 Partly Cloudy	28 Partly Cloudy
29 Partly Cloudy	30 Rainy					

Code:

- Sunny
- Partly Cloudy
- Rainy

Use the calendar above.

On which dates did it rain? _____

How many rain dates were on Sun. _____   Thurs. _____

Mon. _____   Fri. _____

Tues. _____   Sat. _____

Wed. _____

Weekends _____

Weekdays _____

Use your calendar for this month.

Keep a rain record.

Draw raindrops on each day it rains.

Complete the chart.

DAY	DATE	WEEKEND OR WEEKDAY?

Seasons

Name _____

Today is _____

Tomorrow will be _____

Yesterday was _____

There are 4 seasons.

Summer starts on June 21.

Fall starts on September 21.

Winter starts on December 21.

Spring starts on March 21.

When is your birthday?_____What season? _____

Ask your friends when their birthdays are.

Complete the chart.

FRIEND	BIRTHDAY	SEASON

Teacher Notes-Chapter Ten

Calendar Fill In

Objective: Keeping records

Vocabulary: temperature, degrees

Teaching the Lesson: Introduce the calendar for the month using a standard calendar. Help the children to identify the name of the month, the first and last days, holidays, birthdays, and special class days. Duplicate a copy of the blank calendar on page 58 for each child. Help the children to fill in the calendar for this month. Record the temperature for the day. (You may wish to use a class thermometer or a radio report at the same time each day.)

Calendar Detective

Objective: Reading a series of calendars to locate dates

Vocabulary: Veteran's Day, Labor Day, Columbus Day, Thanksgiving Day, Constitution Day, United Nations Day, Halloween, New Year's Eve, Election Day, first, second, third, fourth

Teaching the Lesson: Make a list of holidays, birthdays and special class days which occur in September, October, November and December (the Fall). Help the children to list the dates on which they occur. Discuss the holidays and events with the class. Guide the children to complete the activity on page 42. (You may wish to extend the activity to the calendar for the current year.)

Calendar Crossword Puzzle

Objective: Reviewing calendar information

Vocabulary: cumulative review of days, dates, abbreviations, months, seasons, across, down

Teaching the Lesson: Explain how to find and place information on a crossword puzzle. Help the children to complete the crossword puzzle on page 43.

Birthday Puzzle

Objective: Following directions using calendar vocabulary; mixed practice with computation

Vocabulary: born, remainder, key, year

Teaching the Lesson: Duplicate a list or prepare a chart with the date of birth for each child in the class. Guide the children to determine the day of the week on which they were born by following the directions on page 44. (You may wish to have the children find the day of the week for several members of the class.)

Just for Fun

Chapter Eleven contains a specific activity for every month. Find the activity for this month and duplicate copies for each child.

Calendar Detective

Name _____

Today is _____

Tomorrow will be _____

Yesterday was _____

SEPTEMBER

Monday	Tuesday	Wednesday	Thursday	Friday	Saturday

Be a calendar detective.
What day was Sept. 1 this school year?
Write 1 on the Sept. calendar for that day.
How many days in September? _____
Complete the September calendar.

OCTOBER

Sunday	Monday	Tuesday	Wednesday	Thursday	Friday	Saturday

What day was Oct. 1? _____
How many days in October? _____
Complete the October calendar.

NOVEMBER

Sunday	Monday	Tuesday	Wednesday	Thursday	Friday	Saturday

What day was Nov. 1? _____
How many days in November? _____
Complete the November calendar.

DECEMBER

Sunday	Monday	Tuesday	Wednesday	Thursday	Friday	Saturday

What day was Dec. 1? _____
How many days in December? _____
Complete the December calendar.

The first day of autumn is Sept. 21. Draw a leaf on Sept. 21.
The first day of winter is Dec. 21. Draw a snowflake.

Here is a list of holidays.
Circle each date on the calendar.

Veteran's Day — Nov. 11
U.N. Day — Oct. 24
Thanksgiving — 4th Thurs. in Nov.
Labor Day — 1st Mon. in Sept.
Election Day — 1st Tues. in Nov.
Columbus Day — Oct. 12
Halloween — Oct. 31
New Year's Eve — Dec. 31

Your Days Are Numbered In Calendar Math   © 1980 Cuisenaire Company of America

Calendar Crossword Puzzle

ACROSS

2. There are 12 _ _ _ _ _ _ in a year.

5. Month after September. _ _ _ _ _ _ _

6. Short for Wednesday. _ _ _

8. The first day of summer is in _ _ _ _

9. Short for Monday. _ _ _

10. Another name for fall. _ _ _ _ _ _

11. The season after autumn. _ _ _ _ _ _

13. 12 months = 1 _ _ _ _

DOWN

1. Short for Saturday. _ _ _

2. Short for March. _ _ _

3. The day before Monday. _ _ _ _ _ _

4. The month with Thanksgiving. _ _ _ _ _ _ _ _

6. There are 7 days in one _ _ _ _ .

7. The month after July. _ _ _ _ _ _

8. The first month of the year. _ _ _ _ _ _ _

12. The month after April. _ _ _

Birthday Puzzle

Name_____

Today is _____

Tomorrow will be _____

Yesterday was _____

Find out what day of the week you were born on.
Follow these steps.

When were you born? month _____ day _____ year _____   A._____

  A. Write the last two digits of the year.

  B. Divide that number by 4.
     Ignore the remainder.
     Write the answer   B._____

  C. Write the key number for your birth month.   C._____

January	1	July	0
February	4	August	3
March	4	September	6
April	0	October	1
May	2	November	4
June	5	December	6

  D. Write the date of your birth.   D._____

  E. Add the answers to steps A—D.   E._____

  F. Divide by 7.   F._____

The remainder tells you your birth day.

Sunday	1	Wednesday	4	Friday	6
Monday	2	Thursday	5	Saturday	0
Tuesday	3				

Your Days Are Numbered in Calendar Math   © 1980 Cuisenaire Company of America

January

Name_____

Solve each problem.
Match the answer to a date.
Write the letter in the calendar.
You will get a message.

N $4 \times 4 = $ ___	W $\begin{array}{r} 48 \\ -37 \\ \hline \end{array}$	Y $5 \overline{)100}$	A $10 \div 5 = $ ___	E $\begin{array}{r} 1 \\ 7 \\ +2 \\ \hline \end{array}$
E $5 \times 3 = $ ___	C $\begin{array}{r} 300 \\ -286 \\ \hline \end{array}$	O $\begin{array}{r} 13 \\ \times 2 \\ \hline \end{array}$	U $\begin{array}{r} 3 \\ \times 6 \\ \hline \end{array}$	Y $25 \div 5 = $ ___
P $27 \div 9 = $ ___	O $\begin{array}{r} 5 \\ \times 6 \\ \hline \end{array}$	H $\begin{array}{r} 1 \\ \times 1 \\ \hline \end{array}$	U $\begin{array}{r} 14 \\ 3 \\ 7 \\ +7 \\ \hline \end{array}$	T $5 \times 5 = $ ___
N $3 \times 3 = $ ___	Y $\begin{array}{r} 12 \\ 7 \\ +10 \\ \hline \end{array}$	T $2 \overline{)34}$	R $\begin{array}{r} 534 \\ -515 \\ \hline \end{array}$	P $\begin{array}{r} 2 \\ \times 2 \\ \hline \end{array}$

JANUARY 2001

Sunday	Monday	Tuesday	Wednesday	Thursday	Friday	Saturday
	1	2	3	4	5	6
7	8	9	10	11	12	13
14	15	16	17	18	19	20
21	22	23	24	25	26	27
28	29	30	31			

February Fun

Name _____

Solve each problem.
Find the number in the heart that matches the answer.
Write the letter in the space over the number.
You will get a message.

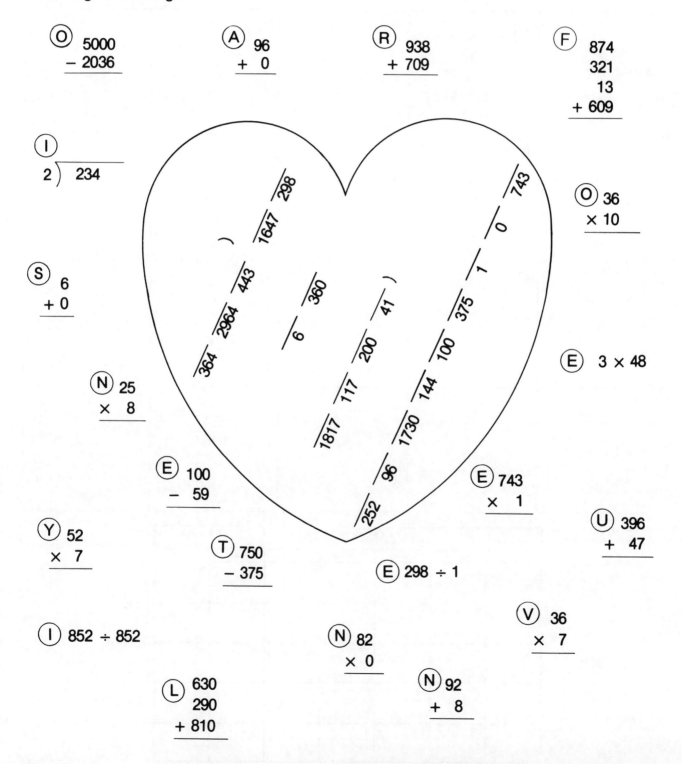

(O) 5000
   − 2036

(A) 96
   + 0

(R) 938
   + 709

(F) 874
     321
      13
   + 609

(I) 2) 234

(O) 36
   × 10

(S) 6
   + 0

(E) 3 × 48

(N) 25
   × 8

(E) 100
   − 59

(E) 743
   ×   1

(U) 396
   + 47

(Y) 52
   × 7

(T) 750
   − 375

(E) 298 ÷ 1

(V) 36
   × 7

(I) 852 ÷ 852

(N) 82
   × 0

(L) 630
     290
   + 810

(N) 92
   + 8

Numbers in heart: 298, 743, 1647, 443, 360, 0, 2964, 41, 1, 375, 364, 6, 200, 100, 743, 117, 144, 1817, 1730, 96, 743, 252

March Maze

Name _____

Some people say, "If March comes in like a lion, it goes out like a lamb."
What do you think that means?

Here's a March maze.
Start at the lion and get to the lamb.
Find the answers to each problem.
Then connect the numbers in the maze in order.

$6 + 5 =$ ____
$2 \times 4 =$ ____
$8 \times 2 =$ ____
$3 \times 6 =$ ____
$7 + 0 =$ ____
$9 \times 1 =$ ____
$7 + 6 =$ ____
$5 \times 8 =$ ____
$8 + 1 =$ ____
$8 \times 7 =$ ____
$7 + 8 =$ ____
$6 \times 3 =$ ____
$5 + 3 =$ ____
$6 \times 2 =$ ____
$9 + 2 =$ ____
$4 \times 1 =$ ____
$6 + 4 =$ ____
$5 \times 5 =$ ____
$9 + 9 =$ ____
$9 \times 5 =$ ____

	11	6	1	16	19
14	8	7	9	13	40
20	16	18	0	42	9
31	21	17	19	15	56
2	19	14	8	18	17
4	18	13	12	9	26
36	24	4	11	5	8
52	30	10	25	34	46
44	17	20	18	45	

April Fool

Name _____

Who's fooling?
Correct the calendar.
How many mistakes can you find?

APIRL 1996						
SUNDAE	MUNDAY	TWOSDAY	WENSDAY	THURSDAY	FRYDAY	SATTERDAY
	1	2	7-5	2×2	5	3×3
15 −8	19 −9	9	20 $\times 5$	6 +7	3 +14	13
7 +7	$5\overline{)75}$	$(2 \times 4) + 8$	$1 + 8 + 9$	18	$(3 \times 5) + 3$	4×5
	11 11 +11	$(8 + 6) \times 2$	$(3 \times 8) + 0$	$(5 + 0) \times 5$	$2\overline{)52}$	$(3 \times 0) \times 9$
4×7	100 −71	(3×5) $+ (3 \times 5)$	31	32		

List your corrections.

_____   _____   _____
_____   _____   _____
_____   _____   _____
_____   _____   _____
_____   _____   _____
_____   _____   _____

Your Days Are Numbered in Calendar Math

May Mistakes

Name_____

Here's a calendar full of mistakes.
Find them and correct them.

			MAY 2104			
SONDAY	MONDIE	TOOSDAY	WENEDSAY	THOORSDAY	FRAYDAY	SATDAY
				1	$4\overline{)8}$	$8\overline{)27}$
$\begin{array}{r}2\\ \times 2\\\hline\end{array}$	$5\overline{)25}$	$\begin{array}{r}3\\ \times 2\\\hline\end{array}$	$\begin{array}{r}7\\ \times 0\\\hline\end{array}$	$\begin{array}{r}8\\ + 1\\\hline\end{array}$	$\begin{array}{r}3\\ \times 3\\\hline\end{array}$	$\begin{array}{r}2\\ \times 5\\\hline\end{array}$
$\begin{array}{r}1\\1\\4\\+ 6\\\hline\end{array}$	$\begin{array}{r}2\\5\\4\\+ 1\\\hline\end{array}$	$3\overline{)39}$	$\begin{array}{r}4\\ \times 7\\\hline\end{array}$	$(2 + 3) \times 3$	$\begin{array}{r}4\\ \times 4\\\hline\end{array}$	$\begin{array}{r}5\\3\\7\\+ 2\\\hline\end{array}$
$\begin{array}{r}320\\ - 3\\\hline\end{array}$	$\begin{array}{r}564\\ - 545\\\hline\end{array}$	$(2 + 3) \times 4$	$(2 \times 10) + 1$	$\begin{array}{r}5\\6\\5\\+ 6\\\hline\end{array}$	$\begin{array}{r}8\\ \times 3\\\hline\end{array}$	$\begin{array}{r}4\\ \times 6\\\hline\end{array}$
$\begin{array}{r}5\\ \times 5\\\hline\end{array}$	$(6+7)\times(1+1)$	$(3 + 0) \times 9$	$\begin{array}{r}7\\ \times 5\\\hline\end{array}$	$(7 \times 4) + 1$	$3\overline{)90}$	$\begin{array}{r}21\\ + 10\\\hline\end{array}$

List your corrections.

_____ _____ _____

_____ _____ _____

_____ _____ _____

_____ _____ _____

_____ _____ _____

_____ _____ _____

June Riddle

Name_____

What's the most important day in June?

To find out, solve each problem.
Find the code letter that matches each answer.
Write it under the answer.
You will get the answer to the riddle.

300 −264	29 +83	306 + 94

75 × 5	606 − 98	116 + 87	6 ×6

97 +58	127 × 4	97 × 0

18 × 8	137 + 0

189 + 14	304 −297	28 × 4	12 ×12	72 +72	500 −125

```
                 ( CODE LETTERS )
     508 - A         137 - F         203 - S
       7 - C         112 - H          36 - T
     155 - D         375 - L           0 - Y
     400 - E         144 - O
```

Your Days Are Numbered in Calendar Math   © 1980 Cuisenaire Company of America

July

Name_____

The Declaration of Independence was signed on July 4, 1776.
That's America's birthday.

Here are the dates when the first 13 states signed the Constitution.
Put them in order along with their code letters.
You will see a birthday message for the U.S.

SCRAMBLED DATES	DATES IN ORDER
(Y) Jan. 9, 1788 Connecticut	(H) Dec. 7, 1787 Delaware
(P) Jan. 2, 1788 Georgia	
(H) Dec. 7, 1787 Delaware	
(P) Dec. 18, 1787 New Jersey	
(A) Dec. 12, 1787 Pennsylvania	

Write the letters in order: H___ ___ ___ ___ ___

SCRAMBLED DATES	DATES IN ORDER
(I) April 28, 1788 Maryland	
(D) July 26, 1788 New York	
(T) June 21, 1788 New Hampshire	
(A) Nov. 21, 1789 North Carolina	
(B) Feb. 6, 1788 Massachusetts	
(H) June 25, 1788 Virginia	
(Y) May 29, 1790 Rhode Island	
(R) May 23, 1788 South Carolina	

Write the letters in order: ___ ___ ___ ___ ___ ___ ___

What is the message?_____

August Treats

Name _____

Play this game by yourself.
Or play with a friend by taking turns.
You will need your calendar for this month and 2 buttons.
Drop the buttons onto the calendar.
Add all the numbers in the boxes the buttons touch.
Get 1¢ for each point.
Use your money to buy treats.
Sometimes you have to save until your next turn.

TURN	NUMBERS	TOTAL	TREAT

   Your Days Are Numbered in Calendar Math   © 1980 Cuisenaire Company of America

September

Name _____

Solve each problem.
Match the answer to a date.
Write the letters on the calendar.
You will get a message.

V	S	R	H	O
9 − 6 3	8 + 7	2 × 4	9 − 8	10 + 9
C	**A**	**H**	**G**	**A**
9 + 7	3 × 2	9 + 8	3 + 4	1 + 1
A	**E**	**R**	**Y**	**O**
5 + 5	3 × 3	13 × 2	13 + 10	9 × 2
E	**A**	**E**	**L**	**T**
2 × 2	5 × 5	8 × 3	4 × 5	3 + 8

SEPTEMBER 1986

Sunday	Monday	Tuesday	Wednesday	Thursday	Friday	Saturday
	1	2	3 ∨	4	5	6
7	8	9	10	11	12	13
14	15	16	17	18	19	20
21	22	23	24	25	26	27
28	29	30				

October

Name_____

Solve each problem.
Color all spaces with an answer that has a 7.
What do you see?

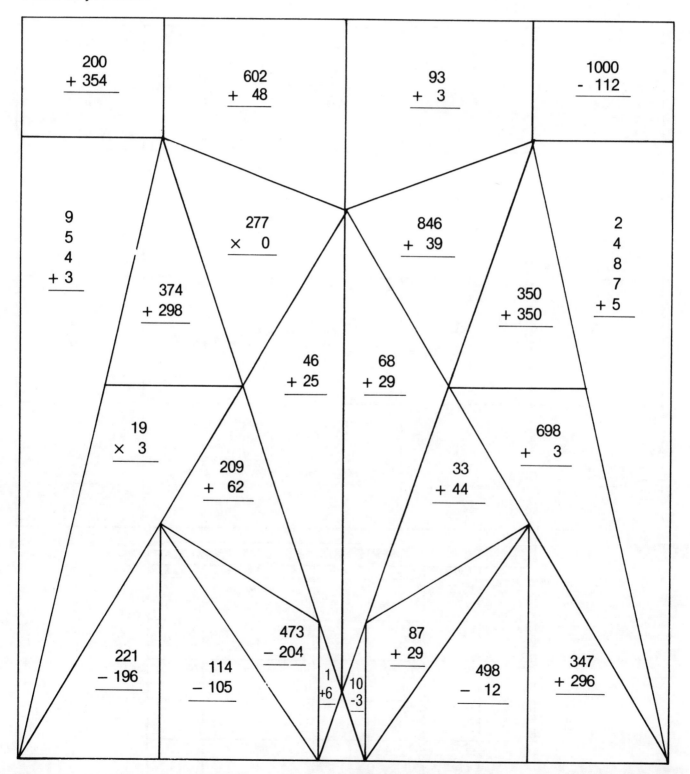

$$\begin{array}{r} 200 \\ + 354 \\ \hline \end{array}$$

$$\begin{array}{r} 602 \\ + 48 \\ \hline \end{array}$$

$$\begin{array}{r} 93 \\ + 3 \\ \hline \end{array}$$

$$\begin{array}{r} 1000 \\ - 112 \\ \hline \end{array}$$

$$\begin{array}{r} 9 \\ 5 \\ 4 \\ + 3 \\ \hline \end{array}$$

$$\begin{array}{r} 277 \\ \times 0 \\ \hline \end{array}$$

$$\begin{array}{r} 846 \\ + 39 \\ \hline \end{array}$$

$$\begin{array}{r} 2 \\ 4 \\ 8 \\ 7 \\ + 5 \\ \hline \end{array}$$

$$\begin{array}{r} 374 \\ + 298 \\ \hline \end{array}$$

$$\begin{array}{r} 350 \\ + 350 \\ \hline \end{array}$$

$$\begin{array}{r} 46 \\ + 25 \\ \hline \end{array}$$

$$\begin{array}{r} 68 \\ + 29 \\ \hline \end{array}$$

$$\begin{array}{r} 19 \\ \times 3 \\ \hline \end{array}$$

$$\begin{array}{r} 698 \\ + 3 \\ \hline \end{array}$$

$$\begin{array}{r} 209 \\ + 62 \\ \hline \end{array}$$

$$\begin{array}{r} 33 \\ + 44 \\ \hline \end{array}$$

$$\begin{array}{r} 221 \\ - 196 \\ \hline \end{array}$$

$$\begin{array}{r} 114 \\ - 105 \\ \hline \end{array}$$

$$\begin{array}{r} 473 \\ - 204 \\ \hline \end{array}$$

$$\begin{array}{r} 1 \\ + 6 \\ \hline \end{array}$$

$$\begin{array}{r} 10 \\ - 3 \\ \hline \end{array}$$

$$\begin{array}{r} 87 \\ + 29 \\ \hline \end{array}$$

$$\begin{array}{r} 498 \\ - 12 \\ \hline \end{array}$$

$$\begin{array}{r} 347 \\ + 296 \\ \hline \end{array}$$

Your Days Are Numbered in Calendar Math © 1980 Cuisenaire Company of America

November Maze

Name_____

Find your way out of the corn fields.
Start at the scarecrow and get to the farmer's barn.
Find the answer to each problem.
Then connect the numbers in the maze in order.

7 + 9 = _____
2 + 3 = _____
3 × 3 = _____
4 + 0 = _____
15 − 8 = _____
3 × 6 = _____
9 × 0 = _____
8 − 0 = _____
11 − 5 = _____
5 × 5 = _____
4 × 7 = _____
8 + 5 = _____
6 × 6 = _____
9 + 3 = _____
8 + 6 = _____
3 × 10 = _____
7 × 7 = _____
6 × 4 = _____
9 + 8 = _____
3 × 4 = _____

December Dreams

Name_____

Play this game by yourself.
Or play with a friend by taking turns.
You will need your calendar for this month and 2 buttons.
Drop the buttons onto the calendar.
Add all the numbers in the boxes the buttons touch.
Get 1¢ for each point.
Use your money to buy the toys.
Sometimes you have to save until your next turn.

TURN	NUMBERS	TOTAL	TOY

The Month of _____

Sunday	Monday	Tuesday	Wednesday	Thursday	Friday	Saturday

The Month of _____

Your Days Are Numbered in Calendar Math © 1980 Cuisenaire Company of America

The Month of _____

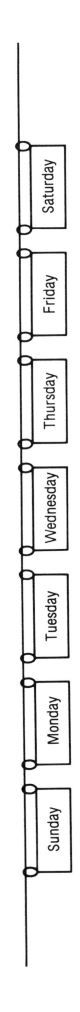

Sunday   Monday   Tuesday   Wednesday   Thursday   Friday   Saturday

The Month of _____

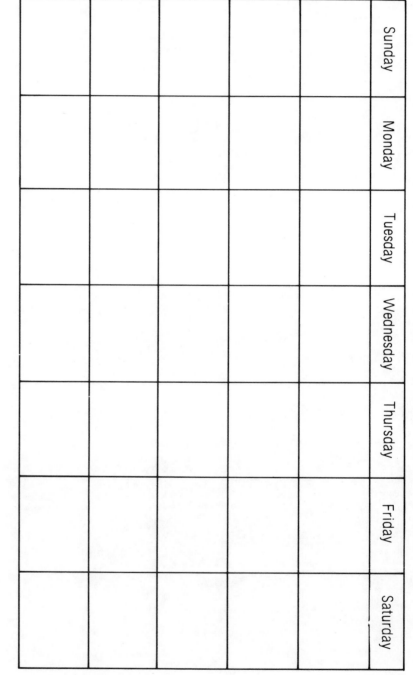

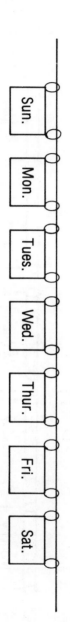

Sun.

Mon.

Tues.

Wed.

Thur.

Fri.

Sat.

The Month of _____

Sunday	Monday	Tuesday	Wednesday	Thursday	Friday	Saturday

The Month of ——————

Sunday	Monday	Tuesday	Wednesday	Thursday	Friday	Saturday

EVENT	DATE	DAY	DAY BEFORE	DAY AFTER

The Month of _____

Sunday	Monday	Tuesday	Wednesday	Thursday	Friday	Saturday

EVENT	DATE	ODD OR EVEN?

The Month of _____

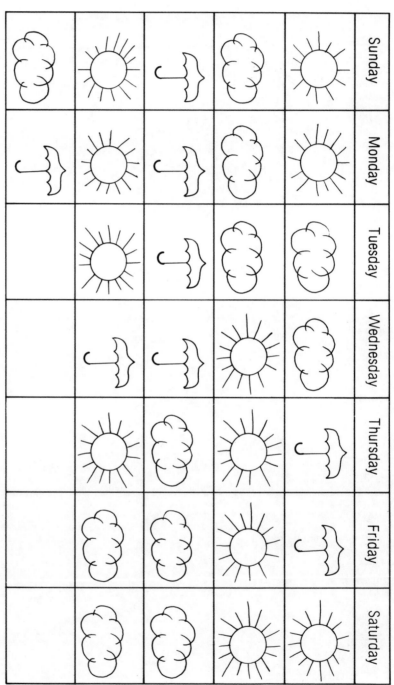

Sunday	Monday	Tuesday	Wednesday	Thursday	Friday	Saturday

CODES

Sunny

Partly Cloudy

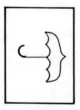

Rainy